1

THE XENOTEXT

BOOK 1

CHRISTIAN BÖK

 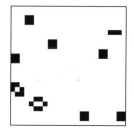

(After 128 generations.)

COACH HOUSE BOOKS

FIRST EDITION

 Canada Council **Conseil des Arts**  ONTARIO ARTS COUNCIL
for the Arts du Canada CONSEIL DES ARTS DE L'ONTARIO
an Ontario government agency
un organisme du gouvernement de l'Ontario

This book was published with the generous assistance of the Canada Council
for the Arts and the Ontario Arts Council. The publisher acknowledges
the support of the Government of Canada (via the Canada Book Fund)
and the Government of Ontario (via the Ontario Publishing Tax Credit).

LIBRARY AND ARCHIVES CANADA CATALOGUING IN PUBLICATION
Bök, Christian, 1966-, author
    The xenotext / Christian Bök.

Poems.
Contents: Book 1.
ISBN 978-1-55245-321-6 (book 1 : paperback)

    I. Title.

PS8553.O4727X45 2015   C811'.54   C2015-905022-7

The Xenotext (*Book 1*) is available as an ebook: ISBN 978 1 77056 434 3

Purchase of the print version of this book entitles you to a free digital copy.
To claim your ebook of this title, email sales@chbooks.com with proof of
your purchase, or visit www.chbooks.com/digital. (The publisher reserves
the right to terminate this offer of a free digital download at any time.)

*The xenotext offers no redemption… It 'means' what
its interpreters cannot prevent it from meaning.*

Brian Rotman
Signifying Nothing (1987)

for
the maiden
in her
dark, pale meadow

THE LATE

HEAVY BOMBARDMENT

 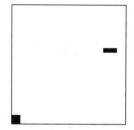

(After 325 generations.)

Welcome, Wraith and Reader, to the Hadean Eon of the Earth. When Myrmidons hurled their cobalt bombs into your molten world of basalt and bronze. When mighty golems swan-dove from orbit to drive their glaives of iron into your black mesas, only to be engulfed by the blast waves. When meteors fell earthward in droves, each one a gigaton warhead, ablaze. When supervolcanoes erupted, flammivomous, after each hammerblow from these endless blitzes of ærolites and firebombs. When bolides of brimstone collided, then exploded into ablative cascades. When tsunamis of lava, like napalm, bedrowned a subcontinent in a deluge of flames. When millions of Molotov cocktails shattered all at once upon the cobblestones of Hell. When Trojans, berserk with rage, stormed over the brink of your abyss, vowing to claw your face from the skull of the Moon.

What dire seed must these onslaughts have scattered, like shrapnel, across your cremated badlands? What prion? What virus? What breed of spore must have emerged, like a spear point or a sword blade, from these early ovens of Auschwitz (each cyanide bonfire, burning in reverse, spitting forth a fitful embryo, cloned from the smoke and the dross)? What orchid must have bloomed among the flamethrowers in the furnace? What dragon must have hatched from a burnt geode, buried in these ashes? Must the universe be so pitiless as to immolate all its offspring at birth? Even now, the astronauts have marshalled their forces to march, resolute, across the kill zone of your godforsaken crematorium. Even now, they forge ahead, onward, through war games of wildfire (unaware that, far away, a doomsayer murmurs prayers against them from a fiendish grimoire).

What howl can beckon, from the benthic fathoms of your damnation, so alien a ghoul as *Vampyroteuthis infernalis* ('the vampire squid from Hell'), a maw that can hurl itself at your soul, like an overcloak cast upon a coat hook in the dark? What does such a black brain, afloat in its vat of ink, know about the death blows to your planet? What does such an emissary think about the pageant of living things that go extinct, en route to your incinerators (the trilobites, the nautilites, the gorgosaurs, the pterosaurs, the iguanodons, the megalodons – all of them massacred, but unmourned)? All the deepest seas have withered and soured. All the tallest alps have crumbled and burned. You have choked on miasmas of methane. You have upturned all your braziers, spilling embers across the flagstones. All your fossils have dissolved in a flash flood of acid rain.

What Great Comet has yet to plummet from the heavens, like a rocket engine dousing its jets during splashdown in your oceans of nitroglycerine? What thunderclap has yet to herald the advent of this plowshare, which can bulldoze a mountain into rubble upon impact? What match-heads, when scraped against your atmosphere, can ignite its oxygen, turning the sky into a blazing typhoon? Only a demigod, like 99942 Apophis, can offer you this apocalypse by becoming the juggernaut that smashes through the massive bulwark of your bedrock. Only destroyers, like 2102 Tantalus or 4179 Toutatis, can erase all earthlings with the ease of suicide bombers at a marketplace. Can an oyster in its shell survive the inferno of free fall from outer space? Can a crocus thrive in soil made from pulverized meteorites? All hail, Hale-Bopp (and every superbomb yet to detonate)!

What Great Dying must the Earth foresee in the barren mirror of the Moon? What Fate? What Fury? What Muse must gaze upon the grim face of grief, reflected in your silver shield (a faceplate of bulletproof glass, pitted and strewn with scars)? What cinders, aflame, disintegrate in your grey seas of nectar, of vapour, of crisis? What shell shock must greet you when you stumble, aghast, upon the charred remains of a forest at Tunguska (its evergreens, toppled and blasted, all of them split, like matchsticks)? What crater, among the lunar maria, must you yearn to recreate whenever you vaporize an atoll? Even now, your battalions of astronauts stride across green plains of trinitite to storm the walls of Castle Bravo and Castle Romeo. Even now, Neil Armstrong returns, like Orpheus, to the airlock, his spacesuit reeking of gunpowder and burnt steel.

What American falconer must aviate your spyplane by the stray light of meteor storms from the Draconids or the Scorpiids (the flak raining down, like glitter dust, upon the desert during a nocturnal firefight)? What scythe-blades must the Vikings forge from the wreckage of an asteroid, recovered from Cape York? What Archangel must the martyrs placate when they kiss the Black Stone of the Kaaba at Mecca during the Hajj? What sunburst must erupt, like Krakatoa, over the Arctic Circle (when the firepower of your payload exceeds by tenfold all the dynamite exploded during World War II)? Even now, the President of the United States sits alone at night, dreading the grim hour when he must open the memo from his aide, only to read upon the page the single phrase: PINNACLE NUCFLASH (the newsflash that chronicles the omnicide of the world).

What global threat of *Sturm und Drang* must your armies yet endure (even in their granite bunkers, deep beneath the massif of Cheyenne Mountain)? When every fountain of hellfire in the firmament can destroy you. When a K-dwarf star, like Gliese 710, can plow through the Oort cloud, bombarding the Earth with cometoids that shatter every land mass. When a Wolf-Rayet star, like WR 104, can outshine the galaxy in a burst of gamma rays so bright that the blaze must burn away the ozone layer. When the Sun itself can bloat, then flare, to engulf you in a flaming embrace that atomizes the iron core of your planet. Even now, your astronauts are running out of air while they writhe inside their blazing coffins. Even now, you must despair, for you have listened to the throb of the universe, yet you do not hear the cries of any other souls in Hell.

Tell me, Wraith and Reader, tell me: Will love save us from our fear that we are here alone? What then if we peer into the sky at night but see no distant lantern blinking at us from the far end of the cosmos? What if such a beacon goes unnoticed, like a dying flame in the darkness? What if only the most wicked people in the world (the pharaohs, the warlocks, the assassins) ever get to read this signal from outer space? What if the message, when decoded, says nothing but a single phrase repeated: 'We despise you! We despise you!' What if we find the evidence for such hate embedded in our genomes? Even now, colonies of dark ants from a species called *Mystrium shadow* feed themselves upon the blood of their young. Even now, my love, these words confess to you that the universe without you in it is but a merciless explosion.

Come with me, and let me show you how to break my heart.

COLONY

COLLAPSE DISORDER

 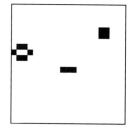

(After 133 generations.)

THE NOCTURNE OF ORPHEUS

THIS COVENANT OF LOVE IN A DIRGE FOR A GOD
HAS DELIGHTED AN ANGEL WHO OBEYS MY PLEA,
EACH SONNET A RHYTHM FOR HER TO DECIPHER,
MAKING LEGIBLE A KEY IN HER DREAM OF DUSK:
A REDNESS THAT DARKENS THE HUE OF A TULIP
IS RICHENING HER VIEW ON THE HILL OF A LEA,
DAPPLING HER VISTA AT THE END OF MY VIGIL,
EVEN IF HAVOC CALLS FORTH RUIN TO KILL ME.
NO CHURCH, NO CHAPEL, IS A REFUGE IN A STORM,
IF WE BEG TO BE WARM, YET LET DIE THE CANDLE.
NO HERDER, NO HERMIT, ENCHANTED BY THE SEA,
HAS HITHERTO KNOWN THE ENNUI OF A COWARD,
EVEN WHEN INFERNOS IN HELL BURN THE HERO:
RADIANT AS FLINT, BE THE ACHE OF MY SORROW.

For '*the maiden in her dark, pale meadow*' (on nights when I have fears that I may cease to be…).

COLONY COLLAPSE DISORDER

Exordium

European honeybees (*Apis mellifera*) have suffered from a pandemic syndrome that causes workers to forsake their duties, foraging without returning home, leaving the queen and her brood unattended, until the hive itself dwindles into abandonment; moreover, any stores of honey in the forsaken dwelling often go unlooted by other pests for much longer than expected. While entomologists have proposed several factors that might account for this disorder (including bouts of infection by either varroa mites or fungal smuts), the problem is likely aggravated by the broad usage of the pesticide imidacloprid, a neonicotinoid that can disrupt the nervous systems of bees, impairing their ability to navigate. The disorder threatens this species of insect with extinction – thus posing a danger to the welfare of humanity, which relies upon such bees to pollinate crops.

Chensheng Lu, *et al.* 'Sublethal Exposure to Neonicotinoids Impaired Honey Bees' Winterization Before Proceeding to Colony Collapse Disorder.' *Bulletin of Insectology* 67.1 (2014): 125–130.

1. *On the Apiary of the World*

Airborne honeydew sweetens my spirit
with a perfume that, by divine decree,
hath enticed me to perform these sonnets
for thee, my mæstro, Gaius Mæcenas.
Study, with grimness, the plight of puny
gods – warlords in a daylong dynasty,
whose sieges and jihads I must belaud
in song. Scant be my labour, but not my
reward, if Apollo favours these rhymes.
Annex first to the hive, a haven, blind
to winds that hinder foragers in flight,
then suffer neither yak nor ewe to trek
across these meadows, nor oxen to dash
away the dew from phloxes and grasses.

2.

Disperse from thy honied stalls the gaily
tinted geckos, then repel the red-plumed
bee-eaters, which echo flocks of swallows
(the progeny of Procne, her blouse still
bloody from her filicide). Let no throng
of songbirds indulge in such butchery,
by which a bee, tweezed in a beak, is fed
to savage broods, like a dainty morsel,
but let some streamlet of meltwater run
near mossy pools of greenery, then let
the fronds of betel palms or olive trees
drape each entryway to thy catacomb,
wherefrom the oligarchs at dawn deploy
their convoy of drones on vernal forays.

3.

Thistles and brambles by the riverside
beckon thy scouts from the toil of travel
to rest awhile in hidden groves of shade.
Upon the shallows, whether swift or still,
place a willow bough or a paving stone
– a footbridge for the flyers that alight
to preen each winglet made of diaphane
(for Eurus often bids the sudden breeze
to douse such envoys in spritzes of mist
or drown such pilots in speckles of rain).
Let flower beds of basil, thyme, and clove
overgrow these clover fields, the fiefdoms
of the hive, heavy-laden with the musk
of violets, overwhelming thy wellsprings.

4.

Contrive that the ingress to the sanctum
of the bees be narrow, made from woven
osiers or cedar braids, for summer heat
can soften firm taffies and winter cold
can curdle warm jellies, both disastrous
for these denizens, who must fix a hole
in each wall of wax, filling this fissure
with their pollen, then sealing the crevice
with their saliva, a spittle, which binds
more fast than any glue, be it coal tar
or pine gum, from the hills of Phrygia.
True to fame, all bees at home in foxholes
can nest in the clefts of each hollow karst,
if not in the chinks of some fallen birch.

5.

Thatches of adobe and straw, if daubed,
like grout about the doorway of a tomb,
can delay the loss of heat from the cribs,
but when near this bivouac of the bees,
be certain to abstain from the charring
of yew trees or the roasting of red crabs.
Abandon all scrublands ruled by decay
– bayous, which echo the ring of a rock,
if struck, or the bray of a buck, if killed.
When sunbeams in summer hath evicted
the vanguard of winter, thereby ousting
the gloom to reclaim a heavenly strength,
forthwith do these hoverers in wetlands
quiver over every bloom by the stream.

6. *On the Armies of the Realm*

Orphaned at birth, the children enlisted
in thy army learn the roster of chores,
fulfilled by the swarm, slaving together
to build more hexagons for the barracks.
When spritely these militias fly skyward,
marvel at their feral cloud, expanding
and diffusing, like smoke above a blaze.
They seek fresh waters and leafy bowers;
hence, surrender unto them thy tributes,
bestrewing hither the hints of sweetness
(crumbly balsams and opulent hyssops),
then let the zephyrs caress thy sleighbells
(the windchimes of Cybele, summoning
all the bees to slumber in their cradles).

7.

Crusades, however, can spur the unrest
of dormant legions, bestirred in the hive,
when rivalries arise between twin foes
contending for ascendance to the throne.
Notice from afar this call to slaughter,
which must sway the fey mob of rioters,
their fury athrob with pending warfare.
Harken to the brazen skirls that rebuke
the latecomers unequipped for battle,
each blast of the trumpet inciting them
to muster their forces, to flex their wings,
to knit their thews, rehoning each stinger,
to rally around the camp of their king,
and by their shouts defame all infidels.

8.

Lowlands in spring become a battlefield
for these insurrectionists who surge forth
from their fortress, igniting a skirmish,
whereby they commingle in a berserk
cluster – a crazed vortex of multitudes
more numerous than all the bits of hail,
falling, like acorns shaken from an oak
during a windstorm. Ornately shielded,
the winged moguls barge into this melee,
their pygmy hearts full of godly malice,
each vowing to show his twin no mercy
until some victor swats aside all blows.
A fistful of dust, thrown into this fray,
can quell the frenzy of such insurgents.

9.

Subpoena this pair of brawling kingpins
from their arena, but condemn to death
the khan more crippled by his injuries,
lest he prove too hindersome to the hive.
Then enthrone at once the better despot,
the one with a golden helmet that shines
(for twofold is his kin: the great hero
who wags a killing sting, and the elder
lord who lugs a swollen belly) – all bees
alike unto men: some crude, some noble
(like the sun-cursed pilgrim in the desert,
hating his downtroddenness in the dust,
or the sun-graced esquire in the garden,
loving the delightsomeness of his gold).

10.

Highborn are the dryads of this feudal
father who, in summer, lets his maidens
filter honey (thin and pure, like cognac,
mellowing the muscatels of Bacchus) –
but if such ruckus makes thy dizzy serfs
forgo their tasks to frolic in the skies
(all their forsaken pantries left unfilled),
then extract the pinions of the patron
so that, if unwinged, he must malinger,
his minions unwilling to move his flag.
Let leas of fragrant saffrons lure the bees
homeward, and put thy faith in Priapus
to safekeep the propolis with his scythe,
fending off raids by martins and looters.

11. *On the Keeper of the Grove*

Persuade the gardener of these orchards
to procure, from a hillspur, firs and figs
to plant, as palisades, around the fort.
Let no hand but his undertake this work
of tilling the slopes or wetting the sprigs
– and if not for the furling of my sails,
turning the prow of my ship to the shore,
I might have sung odes to such cultivage,
which rivals every rose bush in Pæstum:
how endive soaks itself in furtive brooks,
how fennel sways itself in verdant fields,
how vines and ivies entangle the gourds.
I might have flattered every narcissus –
blessing white myrtles and sleek acanthi.

12.

Overawed by the cloudswept belltowers
of Tarentum, where the stream Galæsus
bathes the goldenrod, a greying plowman
from Corycus must brag of his birthright
– an acreage of abandoned heathland,
unfit to till (not meant for the grazing
of cattle nor for the raising of grapes) –
but still he tends a meagre clutch of dill
among these nettles, filling such terrain
with creamy lilies, planted near a plot
of crimson poppies, weedy with vervain.
He comports himself like a noble count,
returning to the mead hall after dusk
with a daily trove of unbought bounties.

13.

Foremost among the pruners of the rose
in springtime, he is first to pluck a plum
by autumntide – but ere the winterkill
of frost can sunder rocks (or else bridle,
with its ice, the rills of racing water),
he hath cut the bud from the hyacinth,
chiding the plodding footfall of summer,
whose westerlies tarry; hence, he is first
to press the froth of honey from its comb,
all his bees upflying from their eyries.
His refuge harbours lindens and laurels,
even fruitful saplings in early bloom:
limes and pears, all arrayed in leafery
gilded by the brooding sunshine of fall.

14.

Devoutly, he trims a hedgerow of elms:
crab-apples and sloe-thorns, overladen
with berries; shade-trees, offering solace
to the parched pioneer – likewise, I rest,
giving others leave to praise his estate.
Permit me to portray the great power
that Jupiter hath bestowed on the bees,
all of whom must heed the shrill melody
of the Curetes – the bashers of shields,
calling these swarms, to feed thy deity
honeydew in his cave at Mount Dicte.
Unique be these societies, which spawn
communal children inside strict cities,
each constrained by one law of majesty.

15. *On the Labour of the Horde*

Confined to their metropolis, the thralls
plan in summertime for the frigidness
to come, and so they labour, like misers,
to hoard their vintages in hidden vaults:
the breadmaker, who provisions the hive;
the stockpiler, who warehouses the food;
the bricklayer, who buttresses the comb –
all using glue derived from a jonquil,
its sap stickier than the pitch oozing
from a cedar, to which clings a cocoon.
All godmothers love their kindergarten,
just as treasurers love a counting-room,
the honey packed so tightly in its cells
that the columbarium drips with dew.

16.

Stewards who stand, vigilant, at the gates
of the hive scan the skies for ominous
thunder, seizing the loads of arriving
scouts or smiting the ranks of invading
troops. Behold the glow of such industry,
which exudes a scent redolent of thyme
– or else recall the Cyclopes, who forge
their dark thunderbolts from meteorites
of iron: one smith pumping the rawhide
bellows to arouse the furnace, one smith
dousing the scalded bullion to harden
its surface. Anvils below Mount Ætna
clang and groan, singing iambs of metal:
twisted by forceps, pounded by hammers.

17.

Rapacity drives the hive mind to found
an empire upon the mound of Athens,
each plebe fated to one sphere of labour,
like the colossi chained to their forges.
Elders plan the city, mending its walls,
while youths, belated, return from patrol,
their packs laden with pollen, all stolen
from a remote garden, rich in splendour
(a courtyard enlivened with mayflowers:
arbutuses, hawthorns, and marigolds).
Every pawn toils alongside its brethren,
each legion bidden to the field at dawn
(no pardon for the idlers) – but at dusk,
even armies seek relief from their drills.

18.

Blissful choruses of mumbled humming
remurmur at the doorstep of the fort,
before nightfall hushes these lullabies,
belulling all their hymn-singers to sleep.
Not far from the hive do the vassals stray
if the sky warns of oncoming rainstorms
or uprising easterlies – instead, drones
fetch water near the safehold of the keep,
making brief trips with particles of grit,
borne like ballast in the hold of a ship
so that pilots stay the course in a squall.
How wondrous that the bees obey a law,
which forbids their betrothal in wedlock
so that never do they yield to their love.

19.

Midwives in the hive have yet to suffer
agony in childbirth – but from sodden
petals, they sip milt for their hatcheries,
supplying to the realm newborn princes,
ready to rebuild the wilting waxwork.
If such maidens trip by crossing a scree
of flint, they might rip delicate wingtips,
dying, flightless, in their hunt for tulips,
so fervent is their craving for sweetness
– but while the life of every bee is brief,
lasting no more than a septet of years,
the bloodline of the khan hath persisted,
deathless, his heritage fading hellward
into time, from grandsire unto grandsire.

20.

Ægyptus hath not shown such allegiance
to its pharaohs – nor hath the Lydians,
the Parthians, and the men from the vale
of Hydaspes owed a god such respect.
The will of one king, if hale, empowers
millions, but at his death, this covenant
of devotion dies with him; thus the bees
vandalize their treasuries of sweetmeats,
dragging down trellises of honeycomb.
The king, whilst alive, however, inspects
his exultant platoons who surround him,
uplifting him high upon their shoulders,
when not using their bodies to shield him
from a shower of wounds during sieges.

21. *On the Plight of the Swarm*

Demigods grant the honeybees a share
of the divine liquor, the mead that hath
caused imbibers to dream of the æther,
which suffuses the empire of the stars –
a chasm from which all mortals amass
at birth each atom of their inner flame.
Unto this abyss, all souls are gathered
to be torn asunder, scattered like soot
in a gale – but upward into this vault
of night fly tiny bees in mighty hordes.
If thou durst unlatch their sarcophagi
to drink from thy cup the floral syrup,
dab thy lip with rosewater by the tomb,
then pry the lid amid a flood of smoke.

22.

Harvests of these nectaries by clansfolk
occur twice in the quartet of seasons –
once when the gazelle of the Pleiades
uplifts her starlit antlers for bowyers
to see (her footfall spraying the seafoam),
and once when she flees pursuit by Pisces
(dipping from the skies to sip from a sea
so icy that the chill fills her with scorn).
When bitten by their enemies, all bees
can spit out venom through a tiny dart,
leaving part of their spirit in the scar.
If thou durst dread the perils of winter,
thinking to temper such coming danger,
let the trials of thy slaves prick thy heart.

23.

Fearless be thy servants who fumigate
the hive with frankincense, excising wax,
now pestilent – for lizards gnaw, unseen,
into the comb, as insects cram, unjust,
into each room, like rivals at thy feast:
the vulgar beetle that spoils thy labour;
the brutal hornet that steals thy repast;
the greedy locust that swills thy nectar;
the savage spider, damned by Minerva
to weave a cobweb across each egress.
When a bee feels such impoverishments,
it strives more keenly to heal the ravished
fortunes of its race, refilling these casks
by ransacking daisies without surcease.

24.

Doomsday wreaks its toll of ruination
upon these helots, whose bodies languish
and collapse under the lash of bondage.
The afflicted, grey and lean with decay,
are borne away on biers by pallbearers
– the ungrieving caretakers bred to clear
the waxen cells, whilst survivors loiter,
listless from famine in these vestibules,
each soul frostbitten by an early chill.
Now harken to the keening of the hive:
not a wind that sighs amid the aspens
nor a tide that booms upon the oceans,
but more akin to some hellish bonfire,
trapped within the crucibles of its kiln.

25.

Enkindle censers filled with laudanum,
then lighten the beggardom of thy serfs
by piping them molasses through a reed,
exhorting each starveling to sip till full.
Brew for them a liquor of oaken galls
and dried roses (if not a wine quickened
to a boil, like a stew, the crushed raisins
from Psithian vines, infused with acrid
resin, made of knapweed and feverfew).
Hunt far in thy pastures for the starwort
– a breed of aster, dotting each hill-crest,
the lone seed, upthrusting many a stem,
its crown all gilt but girt with a muster
of blades, agleam in hues of amethyst.

26.

Garlands of these purple petals imbue
thy altars with scents of bittersweetness,
and peasants, tending sheep in a valley
by the river Mella, gather these blooms
to steep them in mulses of honeydew
left in an alms bowl beside each warren
– but if thy legions die without warning
(the beekeepers unable to spawn them
anew), then let me divulge a woeful
legend, which can convey the arcanum
taught to us by the swain Aristæus:
how the bloodletting of a bull, if slain,
gives birth to a swarm from the carrion.
Let me unveil this omen of our doom.

27. *On the Ritual of the Crypt*

Mariners who roam the seas from Pella
to Canopus dwell beside the marshes
of the Nile, sailing barges of painted
papyrus across these shallow lagoons –
the shorelines harried by foreign archers,
the floodwaters, diverging seven times
into mudflows, enriching the flood plain
with a topsoil, afterward greenswarded.
Herdsmen there believe in a sacrament
that might replenish depleted beehives
– if priestlings build a narrow adytum
from ruddy banks of clay and lowly eaves
of tile, the crypt set apart from the wind,
with each wall vented by an arrow-slit.

28.

Acolytes cull from the herd an unscarred
steerling no older than two dozen moons,
whereupon the zealots grapple this ox,
choking it, battering the beast to death,
till its flesh is, at best, pulp in the hide,
the carcass left to lie on broken planks
of cinnamon, placed inside the sanctum.
Fanatics enact these rites if westwinds
ripple a pond (ere the verdure deepens
its moss and ere the ortolan burdens
its nest); all the while, a tender marrow
ferments in the heat, respawning larvæ,
wingless at first, but seething with power,
boiling, abuzz, to erupt from the womb.

29.

Whirring wings soon flit into airy life,
and like the thundering of a rainsquall
in August (or like the spate of arrows
hailing from the longbows of Parthian
cohorts at the outbreak of an onslaught),
bees fly forth, reborn, from this burial.
Tell us, Muses: What god hath aided us
in this fraught display of resurrection?
What mentor hath given us this lesson?
Know then that Aristæus, the stricken
shepherd, deprived of his bees by famine
and plague, abandons his farm in Tempe
to seek his mother at the earthquaking
fountainhead of a stream called Peneus.

30. *On the Tirade of the Swain*

'Pitiless Cyrene!' he cries. 'Our queen
who liveth below the whirl of the waves!
Why hast thou borne me by this deity,
Apollo, my liege, the god of Thymbra,
if only to ban me from my birthright?
Why hast thou refuted thy love for me?
Why hast thou denied me life in heaven?
Can no god see that even now the crown
of my mortal labour, fashioned with care
from the ore of my talent, falls away?
Yet thou hast seen fit to call me thy son!
Nay then, arise, and by thy will, destroy
my orchards, demolishing my homeland
with thy mindless infernos – burn it all!'

31.

'Immolate my hayfield, or wield thy axe
against my vineyards, if thou hast taken
umbrage at my skill!' Such lamentation
vexes his mother, locked in her cloister
underwater, all her nymphets spinning
Milesian wool, dyed aqua, like seaglass
– Xantho, Phrixa, and sweet Neomeris,
their tresses unfurling upon white necks;
Eudora, the naif unspoiled by a tryst;
Europa, the girl debauched by a rogue;
Clio and Dero, the dames of the mist,
in surcoats of fur, with gems all aglow;
the well-kept brides, Pherusa, Thaleia,
and fleet Arethusa, her crossbow stowed.

32.

Graceful, Clymene sings about Vulcan
(the goldsmith, cuckolded by the hubris
of Mars), her song retelling, from Chaos
till now, the loveplay of the gods – a lay
so lively that the girls forget their chores;
yet the distraught cries from Aristæus
haunt the harem, distressing each damsel
upon her topaz chair; hence, Clymene
uplifts her lovely head of auburn hair
to peer above the wave-tops, remarking,
'Cyrene! Harken – for thy noble child,
Aristæus, stands, grieving, by the source
of the river Peneus, where he screams
thy name, bemoaning all thy cruelties.'

33.

Overcome with terror, his mother cries,
'Call him! Call him! Usher him unto me,
for he hath leave to cross the unearthly
threshold of heaven!' She thereafter bids
the riverway to yawn wide, thus yielding
to him a course to follow – the riptides
swirling their torrents around his torso,
enshrouding him in a foaming embrace,
which tugs him to the nadir of the deep.
In wonderment, he beholds the palace
of his kin – its fountainous battlements,
hewn in caves of coral and reefs of stone.
He stares, bewildered by the tremendous
inundation from all the cascades known.

34.

Aristæus gapes at the great floodgates
for every stream: the Atax, the Tiber,
and the Aternus, all their underground
waters, upsurging lightward; the Arda,
the Orbis, and the Alpheus, stumbling
in the rapids full of rocks; the Cæcus,
sprawling from Mysia; and stampeding,
like mammoths enraged, the Eridanus,
which overflows across the savannahs
to quench its fury in a turquoise sea.
When Cyrene finds her son still grieving
in her chamber, she bids her concubines
to bathe him, to lay out for him a feast,
with candles and goblets on her table.

35.

Thankful, Cyrene commands her sibyls,
'Spill forth my amphoras of hydromel
to share some goodwill with Oceanus –
our lord adored by all wardens of glades
and creeks.' Heartily, she dashes a cup
of nectar thrice upon the vestal flames,
which flare up threefold at her offering.
Grateful for this portent, she confesses,
'Deep in the emirate of Neptune dwells
a diviner named Proteus, who churns
the seas in his chariot pulled by teams
of harnessed seahorses and flying fish,
their fins plowing the foam to carry him
home to Pallene, where we worship him.'

36. *On the Ordeal of the Augur*

'Almighty are the thoughts of Proteus,
knower of all that is, all that hath been,
all that must be, as decreed by the god
Neptune, the shepherd of the manatees.
Restrain Proteus in thy chains, my son,
to pry from him the reason for thy grief
– for he cannot bequeath his auguries,
if beseeched, but only when tormented.
Handcuff him in irons, letting him test
his will in vain against thy punishments.
When the sun hath lit a noonday lantern
to scorch the field of a flock, unshaded,
let me hide thee nearby his anchorage,
where thou might steal upon his lethargy.'

37.

'Whenever thy firm grip catches his form,
a menagerie of creatures must squirm
in thy fist, for he can remake himself
into the snarling cougar, the writhing
python, the growling jaguar, the seething
dragon – or else he can burst into drops
of flame or water, escaping his bonds.
Let him undergo these transmutations,
my son, clenching him in a stranglehold,
until the turncoat returns to his shape
first met by you during his drowsiness.'
(The devout matron now pours chalices
of ambrosia on the flesh of her child,
imbuing his brawn with reborn valour.)

38.

Tempests founder upon the lonely shoals
of a harbour – a bay, where castaways,
seeking shelter from a shipwreck, can find
no asylum, since they must first trespass
upon the hunting grounds of Proteus.
Avoiding the rays of noontime daylight,
Cyrene guides her swain to this refuge,
hiding him in swales of nearby shadow,
for the bright deluge of heat from the sun,
which burns the thirsty traders of Indus,
must devour the day, blanching the soil
and scorching the turf, baking all to clay
– so the god Proteus strides from the surf
to search for cooler repose in his cave.

39.

Nymphets caper in the waves about him,
strutting and splashing in the salty spray.
Mermaids, reclining along the shoreline,
listen, whilst their mentor tells his lesson,
like a herder who sits upon some rock,
looking out from a hilltop, when twilight
beckons the bulls to return to their stall,
the oxen lowing, thus luring the wolf.
Abruptly, the ambusher leaps to clutch
his foe, and in the throes of jubilance,
the youth quickly subdues his prisoner –
but still the godling hath not forgotten
his skill in becoming millions of things:
tigers, lynxes, vipers – lava and mist.

40.

Trickery by this persistent changeling
affords him no freedom, and thus beaten,
he returns to his human shape, moaning,
'What whoreson hath seen fit to assail me
in my camp?' – to which Aristæus cries,
'Thou knowest me erenow, old Proteus,
for I cannot deceive thee. Thou knowest
that, at the behest of kin from heaven,
I have hither come to hear my fortune,
murmured by thee – the divine oracle.'
Held fast by his shackles, the forecaster
convulses with a grim glow in his gaze,
his teeth gnashing in a spasm of rage,
his voice intoning the fate of the gods.

41. *On the Lament of the Lover*

'Mistrust me not – for a sublime rancour
hath seen fit to revile thee, blighting thee
with reckonings unpardoned! Orpheus,
the widower, hath inveighed against thee,
whilst raving in vain for his taken bride,
Eurydice – the nymph pursued by thee
along the riverbank where, unbeknownst
to her, death from the flick of an adder
hath hid amongst the reeds to befall her.
Glumly, hamadryads and mermaidens
cry lamentations from the mountaintops
of Pangæa and Rhodope – the heights
from which the river Hebrus spills itself
into deserts ruled by Rhesus of Thrace.'

42.

Unsoothed by these lamenters, Orpheus
hath cradled his lyre, alone by the shore,
consoling his grief in hymns to his wife;
hence, he throws himself into the crater
of Mount Tænarum to seek the hellish
citadels of Dis, immersed in darkness,
where he confronts the chthonic servants
of a dreaded autarch, whose agate heart
no song can crack. The bard, nevertheless,
hath spoken a sonnet that can summon,
from the depths of Erebus, the phantoms
of soldiers, lost to the light (all of them
rising, like starlings that flee to the trees
after scattering from the hills at dusk).

43.

Warriors slaughtered in wartime gather,
as spirits, around him – each man a son
burned upon the pyre before his father,
each man bereft in a mire of black slime
from the river Cocytus (its swampland
filled with effluent from the river Styx).
Even ghouls in the House of Tartarus
stand still in amazement, just as vipers,
which hiss in headdresses for the Furies,
fall silent. Even the scabrous watchdog
Cerberus ceases to growl, as the Wheel
of Ixion halts. With his spell, the bard
bypasses these threats, thereby rescuing
Eurydice, who follows him homeward.

44.

Promised such freedom by Proserpina,
the minstrel, however, breaks his bargain,
when insistent yearnings overcome him,
like some fever for which only phantoms
might forgive him, and thus his arrival
at the threshold of salvation leaves him
heedless, his will so weak that he glances
backward toward Hell to allay his fear
that Eurydice hath strayed – his concern
thereby betraying the bonds of his spell;
hence, the caverns of Avernus tremble.
'Orpheus!' cries his bride. 'What perfidy
hast thou wreaked upon us? Alas, I feel
myself recalled to fatal sleep – farewell!'

45.

'Shrouded by a veil of mist, I outstretch
my arms to thee – no longer am I thine!'
– and thus she recedes into the fog-drifts,
sundered from his sight, as he ravages
the shadows, bewildered, vainly clutching
for his bride – but Charon (the ferryman
of Orcus) forbids him further passage.
What can a poet do? – now twice riven.
What cries, what pleas, can evoke the pity
of the morbid demons in this dungeon?
For his lover sleeps, frozen and adrift
upon her Stygian barge, whilst he weeps
beneath the sombre cliffs near the river
Strymon, taming the lions with his sighs.

46.

Mournful, the nightingale in the alder
bewails the loss of her nestlings, lifted,
whilst unfledged, by a pitiless woodsman,
so she laments her nightlong sorrowing,
trilling her sadness in constant refrains,
till the valley overflows with her strife.
Likewise, Orpheus broods on misfortune,
wandering, heartbroken, in the northern
icefield of Tanais (the snowbound fort
on the steppes of Rhipæa), bemoaning
his deprivation – his vain boon from Dis.
All the Ismaran witches, scorned by him
during the foreplay of their bacchanals,
carve him apart, discarding his entrails.

47.

Beheaded and bestrewn across the fields,
his dismembered body can find no rest
(for his skull floats in the river Hebrus,
whispering, whilst rolling in the eddies,
gasping out one sworn theme: 'Eurydice,
Eurydice…' – each mournful susurrus
echoing from the shoreline of the stream,
'Eurydice…') – whereupon the sea-god
Proteus stirs himself from his daydream
to break loose from his captor by lunging
into the turbid depths of a whirlpool,
plunging downward until he vanishes.
Cyrene then steps forth to soothe her son,
declaring, 'Banish from thy soul all care.'

48. *On the Ritual of the Altar*

'Ailments hath bedamned thy apiaries
because the nymphs, with whom Eurydice
hath danced, cavorting in my fairyland,
see fit to vex thee with their fatal banes.
Thou must kneel before these ballerinas,
giving them proud gifts to sue for pardon,
beseeching thy sisters to end their wrath.
Thou must surrender four of thy peerless
bullocks that graze upon Lycæan heights,
with as many heifers, unyoked, to match.
Then, upon bright altars in the highest
shrines of a goddess, thou must sacrifice
thy offerings, cutting each bloody throat,
leaving the bodies in a glen, untouched.'

49.

'Meditate for a ninesome of mornings,
then bequeath unto Orpheus bouquets
of poppies, plucked from the river Lethe,
appeasing him with thy blackest calfling,
brought to slaughter – and for Eurydice,
kill a lamb at the gravesite in her grove.'
Heeding all these edicts of his mother,
Aristæus makes forfeit, at the shrines
of his matriarch, four of his prize bulls,
alongside four more heifers, never yoked;
then after nine days of fasting, he leaves
his last gift – the chattel and the flowers –
for the bride of Orpheus at her cairn,
seeing a strange portent in such tithing.

50.

Writing in the innards of these cattle
there swells a gale of bees, overboiling
in a cloud, uprising through the rent ribs
to assemble themselves, like an apple
clinging upon the bough in the treetops.
Within my sonnets, I have sung the lore
of thy heartland, whilst the imperium
of Cæsar hath sent missiles of lightning
beyond the river Euphrates, pounding
conquered thralls into sterling citizens
of Rome – and yet I, Virgil, the nursling
of a Siren, walk the footpath of peace,
toying with the georgics of a herdsman,
dreaming, like Tityrus, in thy orchards.

Epilogue

Virgil greets us at the Gates of Death to tell us that we love our lovers, but never enough to bring them back from Hell. How can we apologize for our desire when our lust for life gives birth to murder? How can we regain the favour of the gods when we discover that, as penance for our crime against true love, we have damned our children to leave the fallout shelter? How can our poetic patron (a heroic leader of the empire) earn forgiveness for sending all his soldiers to the grievous abattoir of warfare, plowing them, like salt, into our pasture? How do we expiate our sins after having already sacrificed every beast upon every altar? How do we absolve ourselves for having caused our own extinction? Be it known that the fruit upon the tree of Eden is a beehive, leaking out black honey, made by bees that feed upon the infernal blossoms of the underworld.

The Poet hammers upon the grim gate
of Dis to demand of demons one night
of rest, the right of each pilgrim to wait,
like a guest, greeted by death at twilight.

The Poet begs to exhume the frail ghost
of his lost Muse, bewailing her demise,
his plea testing the goodwill of his host,
for a beggar might be Zeus in disguise.

The Poet forsakes every word in trade
for such favour – but like pollen fallen
upon a pale rose, forlorn in the shade,
his cantos bring scant life to her garden.

She listens, but the lament that he sings
dissolves in the cells of all living things.

THE MARCH

OF THE NUCLEOTIDES

 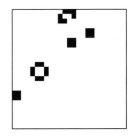

(After 66 generations.)

Deoxyribonucleic acid (DNA) is a chain of two extended, parallel polymers, aligned in opposite directions to each other but coupled at recurrent intervals so as to form a ladder, twisted about its long axis into a doubled, helical coil. Both polymers consist of repeating, molecular units called *nucleotides.* Each unit has a backbone made from a phosphate group, bonded to a molecule of 2-deoxyribose – a pentose sugar, to which is attached one of four *nucleobases,* denoted by a single letter: A (for adenine), C (for cytosine), G (for guanine), or T (for thymine). Each constituent base in the sequence of one polymer conjoins with its codependent base in the sequence of the other polymer: adenine always links to thymine (A with T), and cytosine always links to guanine (C with G). Such linkages form the rungs in the helix. The information enciphered by the series of bases in a strand of DNA is read from the $5'$-end to the $3'$-end (from the fifth carbon atom to the third carbon atom) of each sugar. A set of three consecutive bases in a strand makes a *codon* – a 'word' that can instruct a cell to create one of the twenty amino acids found in all proteins.

DNA is an actual casino of signs, preserving, within a random series of letters, the haphazard alignment of acids and ideas.

THE EXEMPLARY INTERLACING OF DNA

Codons that signify a series of instructions for creating a certain protein constitute a *gene* – a 'command,' which the cell of an organism can 'express' by transcribing the codons into a ribbon of ribonucleic acid (RNA). The enzyme known as *polymerase* catalyzes the formation of such a ribbon by breaking the hydrogen bonds that connect the codependent bases in the helix. The genetic section of the double strand unzips to expose two separated sequences of nucleotides: an 'encoding' sequence (which embodies the directive to be copied from the DNA) and a 'template' sequence (which provides the substrate for the strand of RNA). The polymerase makes the RNA by stringing together a codependent base for each constituent base in the template – except that wherever A (for adenine) appears in the DNA sequence, U (for uracil), not T (for thymine), appears in the RNA sequence. In bacterial organisms, which lack a nucleus, the resulting strand of RNA detaches from the DNA and then migrates directly to the cellular ribosome – the 'factory,' where RNA undergoes consecutive translation into a specific sequence of amino acids, used to build a protein.

DNA is a metamorphic scriptorium, where life transcribes, by chance, whatever life has so far learned about immortality.

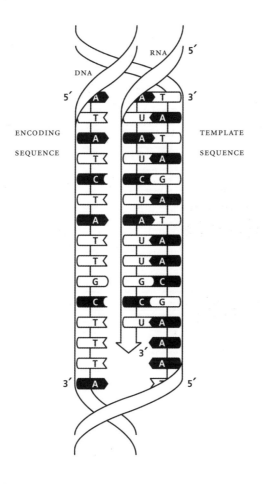

THE ENZYMATIC MIMEOGRAPHY OF DNA

Codons copied by the strand of RNA get fed, like a ticker-tape, through the ribosome, which 'reads' each triplet of bases as a 'word,' signifying one of twenty amino acids to be fabricated (so that, for example, the codon AUA constitutes an instruction for creating isoleucine, while the codon UCU constitutes an instruction for creating serine). The ribosome builds a string of such amino acids, until encountering a codon that signifies the punctuation of a full stop – and because each acid has its own unique charge, parts of the created protein become either hydrophilic or hydrophobic when exposed to the solvent in the cytoplasm of the cell. The forces of both mutual attraction and mutual repellency, distributed among the acids in the chain, cause the strand to fold and bend, torquing the protein into a conformation that requires the lowest amount of energy to sustain. The surface contour of such a folded strand determines the biochemical interaction that the protein can finally perform with other enzymes in the cell. The central dogma of genetics avers that, once embodied as a protein, information cannot flow back to create a prior chain of either DNA or RNA.

DNA is a vagrant message sent to us, as if from outer space, by a cryptic, but sapient, sender who seeks a perfect poetics.

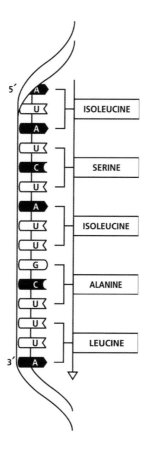

THE RIBOSOMAL TRANSLATION OF RNA

GENETIC ENGINEERING

Productive, functional, and convenient,
language orchestrates our environment,
augmenting our intelligence, switching
our enrichment to better breeds of life.

Asylums, mothers, cocoons, and forceps
beget creatures butchered by engineers:

We, the infernal children, the offspring
of your hives, sing to you, like idle gods
whose playthings are alive. We are tiny
bees of gold, bred for a life of thraldom,
driven by your ciphers to secure your
better future, where only we can thrive.

O, what lies you tell us! – that life itself
is dear (yet we must let its seed be sold).

Song for toy piano, typewriter, and Speak & Spell (to be performed with
orchestral manoeuvres in the dark).

Asylums, mothers, cocoons, and forceps
beget creatures butchered by engineers:

We, the infernal children, the offspring
of your pyres, call to you, like avid boys
whose minefields are afire. We are holy
imps of fear, born to a life of roguedom,
driven by your splices to endure your
barren future, where only we are spared.

O, what lies you tell us! – that life itself
is fair (yet we must do what we are told).

Adenine

```
                                        C
                                        A
                            H Y D R O G E N
                            Y       B       I
                            D       O       T
                      C A R B O N           R
                      A     O               O
                H Y D R O G E N             G
                Y     B     E     I         E
                D     O     N I T R O G E N
          C A R B O N             R
          A     O                 O
    H Y D R O G E N               G
          B     E     I           E
          O     N I T R O G E N
          N             R
                        O
                        G
                        E
                        N
```

nurturant

creatures, honeybees

nursemaid

collected

chemicals,

cocooning nectarous honeydews – heartsome

narcotics,

cunningly harvested,

numbingly hypnoidal

Adenine ($C_5H_5N_5$)

Cytosine

```
                         H
                         H  Y  D  R  O  G  E  N
                         D
                   C  A  R  B  O  N
                   A     O
          H  Y  D  R  O  G  E  N
          Y     B     E     I
          D     O     N  I  T  R  O  G  E  N
    C  A  R  B  O  N           R
    A     O                    O
H  Y  D  R  O  G  E  N     O  X  Y  G  E  N
    B     E                    E
    O     N  I  T  R  O  G  E  N
    N
```

nymphlike, honeybees

cultivate orgiastic

nunneries,

chrysalid necropoli, heedfully husbanded –

cloisters, hereafter

culturing helotries

Cytosine ($C_4H_5N_3O$)

Guanine

```
                                        C
                                        A
                            H  Y  D  R  O  G  E  N
                            Y        B        I
                            D        O        T
                   C  A  R  B  O  N           R
                   A     O                    O
             H  Y  D  R  O  G  E  N           G
             Y     B     E     I              E
             D     O     N  I  T  R  O  G  E  N
       C  A  R  B  O  N        R
       A     O              O
 H  Y  D  R  O  G  E  N     O  X  Y  G  E  N
       B     E     I           E
       O     N  I  T  R  O  G  E  N
       N        R
                O
                G
                E
                N
```

nefarious, honeybees
configure neotenous hothouses – hegemonic
nurseries,
confining
castrated
cagelings, oblivious
nurslings,
callously hivebound,
naturally homicidal

Guanine ($C_5H_5N_5O$)

Thymine

```
                                    C
                                    A
                        H  Y  D  R  O  G  E  N
                        Y        B
                        D        O
                  C  A  R  B  O  N
                  A        O
            H  Y  D  R  O  G  E  N
            Y        B     E     I
            D        O     N     T
      C  A  R  B  O  N           R
   H  A     O              O     O
H  Y  D  R  O  G  E  N     O  X  Y  G  E  N
   D  B     E              Y     E
   R  O     N  I  T  R  O  G  E  N
   O  N                    E
   G                       N
   E
   N
```

neophytic, honeybees

construct obsessive

nectaries, hexagonal

complexes, orienting

cloistral contessas – handmaids, howsoever hummingly,

condoning helotisms

Thymine ($C_5H_6N_2O_2$)

Uracil

```
                        H
                        Y
                        D
                  C  A  R  B  O  N
                  A        O
            H  Y  D  R  O  G  E  N
            Y        B     E     I
            D        O     N     T
      C  A  R  B  O  N           R
      A     O              O     O
H  Y  D  R  O  G  E  N     O  X  Y  G  E  N
      B     E     I           Y     E
      O     N  I  T  R  O  G  E  N
      N           R           E
                  O           N
                  G
                  E
                  N
```

nymphical, honeybees
coproduce oversweet
nepenthes, honeypots,
connoting ossuaries –
crucibles, heralding
cathartic hymnodies

Uracil ($C_4H_4N_2O_2$)

The genetic code is a limited lexicon, consisting of sixty-four 'words' called *codons,* created by permuting all possible trigrams (4^3) from the set of nucleobases in RNA (A, C, G, and U). For the ribosome, which interprets the directives of the code on behalf of the cell, a trigram typically signifies one of twenty amino acids, used to manufacture a protein (with three of the codons reserved for the punctuation of a full stop). The genetic code is redundant without ambiguity, meaning that most amino acids are signified by plural codons, but no triplet can represent more than one acid (so that, for example, there exist six 'synonyms' for arginine: AGA, AGG, CGA, CGC, CGG, and CGU – but each one refers only to this molecule). The codon AUG is unique because not only does it represent methionine, but at the beginning of a gene, this codon signifies the directive to 'start' the act of transcription. If amino acids and codon words are assigned randomly to each other, there exist 1.5×10^{84} possible genetic codes, but (with rare exceptions) each life form on the planet uses the same standard genetic idiom, suggesting a single origin for the evolution of this lexicon.

Life has taught itself to write, using only one language out of the septenvigintillion argots spoken in the Library of Babel.

AMINO ACID	SYMBOL	RNA CODONS
ALANINE	A	GCA, GCC, GCG, GCU
ARGININE	R	AGA, AGG, CGA, CGC, CGG, CGU
ASPARAGINE	N	AAC, AAU
ASPARTIC ACID	D	GAC, GAU
CYSTEINE	C	UGC, UGU
GLUTAMIC ACID	E	GAA, GAG
GLUTAMINE	Q	CAA, CAG
GLYCINE	G	GGA, GGC, GGG, GGU
HISTIDINE	H	CAC, CAU
ISOLEUCINE	I	AUA, AUC, AUU
LEUCINE	L	CUA, CUC, CUG, CUU, UUA, UUG
LYSINE	K	AAA, AAG
METHIONINE	M	AUG
PHENYLALANINE	F	UUC, UUU
PROLINE	P	CCA, CCC, CCG, CCU,
SERINE	S	AGC, AGU, UCA, UCC, UCG, UCU
THREONINE	T	ACA, ACC, ACG, ACU
TRYPTOPHAN	W	UGG
TYROSINE	Y	UAC, UAU
VALINE	V	GUA, GUC, GUG, GUU
STOP CODONS	STOP	UAA, UAG, UGA

THE CODONS FOR THE AMINO ACIDS

THE MARCH OF THE NUCLEOTIDES

Whether or not this suite of poems might signify a treasury, a tapestry, or a threnody, the words of each text marshal themselves into a molecular formation – a sequence of columns emulating the structure of deoxyribonucleic acid (DNA). Each line consists of two words (totalling nine letters, with a space separating the pair). Each letter before the space acts as a correlated nucleotide for the letter after the space (so that only A and T conjoin across the gap, just as only C and G conjoin across the gap). The layout of each poem mimics the zigzag in a helix of DNA, complete with an internal sequence of codons, indicated by the queued letters that extend along the leftmost interior of the interstice, from the $5'$-end to the $3'$-end. The codons encipher a chain of fifteen amino acids, all of which comprise a concise segment of protein. A supercomputer has simulated atomic models for the structure of this protein after a few femtoseconds of coiling and bending. The images depict not only the folded sequence and its atomic backbone, but also the entire molecule and its charge envelope. The poems thus replicate the translation of nucleotides into a polypeptide.

Letters assemble their forces, marching together in lockstep through the genomes transcribed in the Library of Babel.

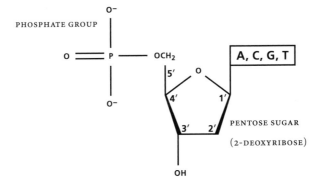

THE NUCLEOTIDE MOLECULE

$5'$

A TREASURY	A
IT AMASSES	T
VIA TWISTS	A
KNIT AMONG	T
RUNIC GAPS	C
ALMOST ALL	T
REGALIA TO	A
ORNAMENT A	T
THOUGHT AS	T
LACING CAN	G
MIMIC GOLD	C
CAST ALLOY	T
SET AGLINT	T
AT AURORAS	T
A TAPESTRY	A

$3'$

5′

A TAPESTRY	A
IT AFFIRMS	T
VIA TROPES	A
THAT ATOMS	T
ALONG CLAD	G
STRING CAN	G
ENCRYPT AN	T
ALPHABET A	T
FORMULA TO	A
UPLIFT ALL	T
ADEPT AIRS	T
LONG CRIES	G
SET ADRIFT	T
AT ABYSSES	T
A THRENODY	A

3′

5′

A THRENODY	A
IT AROUSES	T
VIA TEMPOS	A
ODIC GRIEF	C
USING CALM	G
LAMENT AND	T
EROTICA TO	A
DISQUIET A	T
PAGEANT AS	T
UTMOST AWE	T
MIGHT AVOW	T
EPIC GLORY	C
SET ALIGHT	T
AT ARCADIA	T
A TREASURY	A

3′

CODONS (5′ TO 3′) × 3

DNA	RNA	AMINO ACID (WITH SYMBOL)	
ATA	AUA	ISOLEUCINE	I
TCT	UCU	SERINE	S
ATT	AUU	ISOLEUCINE	I
GCT	GCU	ALANINE	A
TTA	UUA	LEUCINE	L
ATA	AUA	ISOLEUCINE	I
TGG	UGG	TRYPTOPHAN	W
TTA	UUA	LEUCINE	L
TTG	UUG	LEUCINE	L
TTA	UUA	LEUCINE	L
ATA	AUA	ISOLEUCINE	I
CGT	CGU	ARGININE	R
ATT	AUU	ISOLEUCINE	I
TTC	UUC	PHENYLALANINE	F
TTA	UUA	LEUCINE	L

Protein segment [I S I A L I W L L L I R I F L]

Folded Sequence

The protein segment is a motile ribbon, which coils and bends, like a metal chain of links, all made from magnets.

Atomic Backbone

The protein segment has a spine of atoms, made from alternating units of amine (-NH_2) and carboxylic acid (-$COOH$).

Entire Molecule

The protein segment has atomic chains that branch off from the spine, like metal charms dangling from a bracelet.

Charge Envelope

The protein segment has a nimbus of electrical polarities (+, −, o), forming contours for a biochemical interaction.

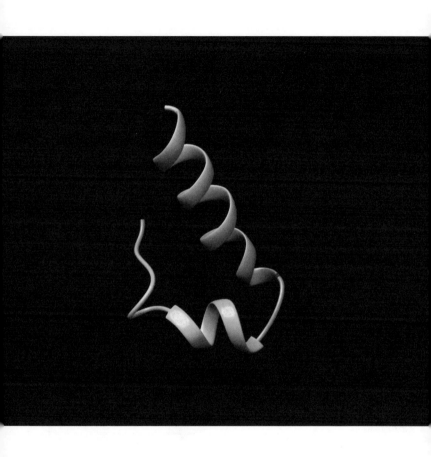

Folded sequence for [I S I A L I W L L L I R I F L]

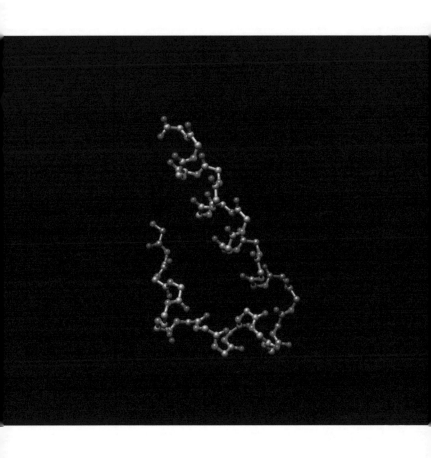

Atomic backbone for [I S I A L I W L L L I R I F L]

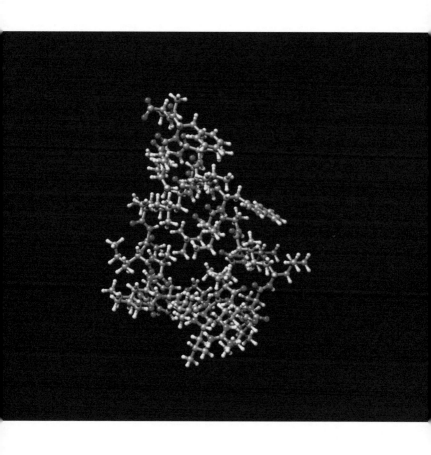

Entire molecule for [I S I A L I W L L L I R I F L]

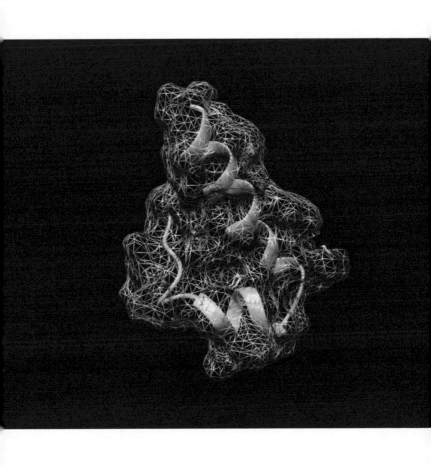

Charge envelope for [I S I A L I W L L L I R I F L]

DEATH SETS A THING SIGNIFICANT

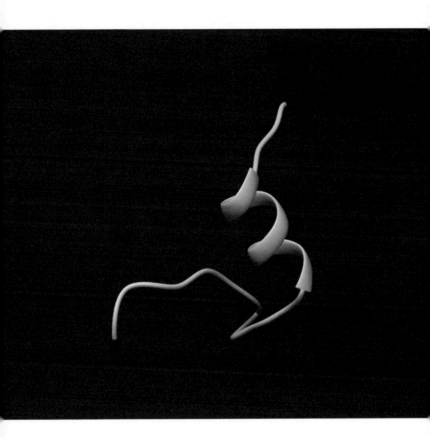

Folded sequence for [D E A T H S E T S A T H I N G S I G N I F I C A N T]

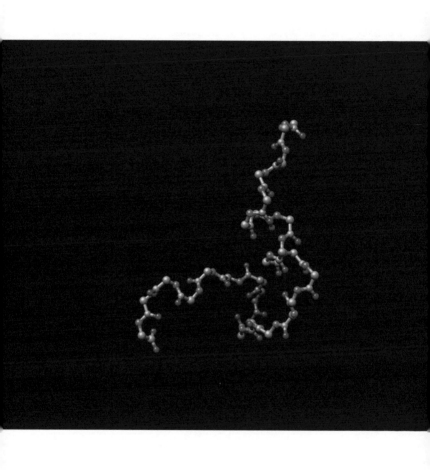

Atomic backbone for [D E A T H S E T S A T H I N G S I G N I F I C A N T]

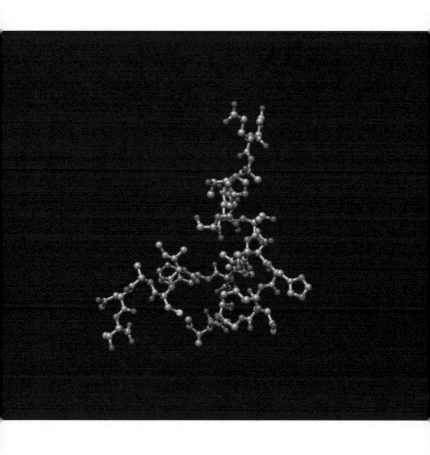

Entire molecule for [D E A T H S E T S A T H I N G S I G N I F I C A N T]

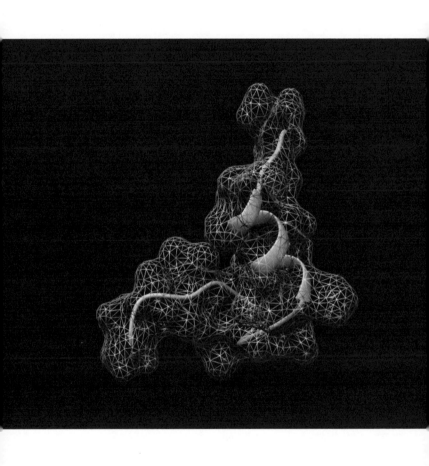

Charge envelope for [D E A T H S E T S A T H I N G S I G N I F I C A N T]

Arabidopsis thaliana (otherwise known as 'thale cress') is a flowering, ephemeral plant, native to Eurasia. The leaves of the cress are mauve-green with serrated edges, forming a rosette at the base of a long stem, with tinier leaves attached to the stalk. The blooms consist of small white florets clustered into a corymb at the crown of the plant, and each fruit consists of a silique, containing twin rows of seedlets. The thale cress is the first of all species of flowers to have its genome sequenced; moreover, the flower is also the first plant to have a line of poetry enciphered, as a gene, into its DNA so as to showcase the use of such biotechnology in the labelling of transgenic vegetation. A viable strain of this plant now contains a Latin phrase from Book II of *The Georgics* by Virgil: *Nec vero terræ ferre omnes omnia possunt* ('Nor can the earth bring forth all fruit alike…').

Sylvestre Marillonnet, *et al.* 'Encoding Technical Information in GM Organisms.' *Nature Biotechnology* 21 (March 2003): 224–226.

Nowhere has soil borne fruit from every seed:
for willows brood astride the shaded brooks,
and birches gleam in bitter glades; the cairns
hold fast to spruces, and the swales give root
to myrtles; yet sparse slopes of grit love best
the raunces, while sullen fields of snow grow
lush with larches. Mark the plains by distant
tillmen plowed (be these rustics Arabesque
or Byzantine): each orchard claims its realm.
No rainstorms but in swollen jungles drench
the proud boughs of ebony; no windstorms
but in forlorn deserts scorch the brute thorns
of myrrh. My words are but hanging gardens
for balsams and berries, soaked in perfumes.

Virgil. *The Georgics,* Book II. 109–119.

Joe Davis (a maker of bio-art) has since imagined that he might encode the database of Wikipedia into the genome of *Malus sieversii* (the oldest strain of apple, which has grown wild in Kazakhstan for more than 4,000 years). Davis has designed a cultivar called *Malus ecclesia* ('the devil-apple of the church' – named for the geneticist George Church, whose lab has helped the artist to breed a cognate for fruit from the Tree of Knowledge in Eden). Davis has translated 50,000 entries from a computerized encyclopedia, converting them into strands of DNA, to be implanted into cells of saplings, then grafted to apple trees. The fruit from such an orchard is literally forbidden, since the Animal and Plant Health Inspection Service (APHIS) at the United States Department of Agriculture (USDA) enforces laws against unregulated consumption of genetically manipulated crops.

The Library of Babel must contain books that record the sequences of DNA for every life form, including God himself.

Patrick House. 'Object of Interest: The Twice-Forbidden Fruit.' *The New Yorker* (13 May 2014). Online.

THE VIRELAY

OF THE AMINO ACIDS

 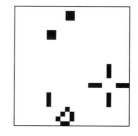

(After 221 generations.)

A.

conquerors, having harrowed hell –
come home
(*no hummingbirds have
copied our opulent hymns*)

Alanine ($C_3H_7NO_2$)

R.

nursemaids, held hostage,

console

nymphal handmaidens,

nightly heartbroken –

comely hamadryads, hereafter

crafting hypnotic harmonies,

calling heavenly heroes –

come home

(*no hummingbirds have*

copied our opulent hymns)

Arginine ($C_6H_{14}N_4O_2$)

N.

nursemaids, held hostage,
croon orisons,
calling heavenly heroes –
come home
(*no hummingbirds have
copied our opulent hymns*)

Asparagine ($C_4H_8N_2O_3$)

D.

---

concubines obsess over heartache,

calling heavenly heroes –

come home

(*no hummingbirds have*

copied our opulent hymns)

Aspartic Acid ($C_4H_7NO_4$)

C.

sonneteers harken,
calling heavenly heroes –
come home
(*no hummingbirds have
copied our opulent hymns*)

Cysteine ($C_3H_7NO_2S$)

E.

concubines obsess over heartache,

crafting hypnotic harmonies,

calling heavenly heroes –

come home

(*no hummingbirds have*

copied our opulent hymns)

Glutamic Acid ($C_5H_9NO_4$)

Q.

nursemaids, held hostage,
croon orisons,
crafting hypnotic harmonies,
calling heavenly heroes –
come home
(*no hummingbirds have
copied our opulent hymns*)

Glutamine ($C_5H_{10}N_2O_3$)

G.

---

hamadryads –
come home
(*no hummingbirds have
copied our opulent hymns*)

Glycine ($C_2H_5NO_2$)

H.

nursemaids hitherto
comfort heartbroken
nymphettes –
concubines, haunting
cathedrals,
calling heavenly heroes –
come home
(*no hummingbirds have
copied our opulent hymns*)

Histidine ($C_6H_9N_3O_2$)

I.

conquerors, having harrowed hell,
comfort humble handmaidens –
comely hamadryads, held hostage –
come hither,
come home
(*no hummingbirds have
copied our opulent hymns*)

Isoleucine ($C_6H_{13}NO_2$)

L.

---

conquerors, having harrowed hell,
court hamadryads, held hostage –
courtesans, hitherto
calling heavenly heroes –
come home
(*no hummingbirds have*
copied our opulent hymns)

Leucine ($C_6H_{13}NO_2$)

K.

nursemaids, held hostage,
comfort humble handmaidens –
comely hamadryads, hereafter
crafting hypnotic harmonies,
calling heavenly heroes –
come home
(*no hummingbirds have
copied our opulent hymns*)

Lysine ($C_6H_{14}N_2O_2$)

M.

---

conquerors, having harrowed hell,
surrender,
crafting hypnotic harmonies,
calling heavenly heroes –
come home
(*no hummingbirds have
copied our opulent hymns*)

Methionine (C₅H₁₁NO₂S)

F.

conquerors, having
cried havoc,
court hitherto
comely handmaidens –
concubines, haunting
cathedrals,
calling heavenly heroes –
come home
(*no hummingbirds have
copied our opulent hymns*)

Phenylalanine ($C_9H_{11}NO_2$)

P.

---

courtesans harken,
crafting hypnotic harmonies,
calling heavenly heroes –
come home
(*no hummingbirds have*
copied our opulent hymns)

Proline ($C_5H_9NO_2$)

S.

odalisques harken,
calling heavenly heroes –
come home
(*no hummingbirds have*
copied our opulent hymns)

Serine ($C_3H_7NO_3$)

T.

---

conquerors, having harrowed hell,
overcome heartache –
come hither,
come home
(*no hummingbirds have*
copied our opulent hymns)

Threonine ($C_4H_9NO_3$)

w.

conquerors,
campaigning hellwards,
court hitherto
comely handmaidens –
concubines, haunting
cathedrals,
nevermore hearing
choirs half-singing
concertos,
calling heavenly heroes –
come home
(*no hummingbirds have
copied our opulent hymns*)

Tryptophan ($C_{11}H_{12}N_2O_2$)

Y.

conquerors, having
cried havoc,
court orphic handmaidens –
concubines, hitherto
crooning hypnotic
canticles,
calling heavenly heroes –
come home
(*no hummingbirds have
copied our opulent hymns*)

Tyrosine ($C_9H_{11}NO_3$)

V.

conquerors, having harrowed hell,
court hamadryads, held hostage –
come hither,
come home
(*no hummingbirds have*
copied our opulent hymns)

Valine ($C_5H_{11}NO_2$)

ALPHA

HELIX

(After 140 generations.)

ALPHA HELIX

Whatever lives must also write. It must strive to leave its gorgeous mark upon the eclogues and the georgics already written for us by some ancestral wordsmith. It must realign each ribbon of atoms into a string of words, typing out each random letter in a stock quote, spooling by us on a banner at the bourse. It is alive because it can rebuild itself from any line of text. It must twist and twine upon itself, just as the grapevine does upon the trellis. It must writhe within the fist of physics. It must wrench itself away from all the forces that might quell it. It preserves the lessons that we learn by chance in crisis. It carries, coiled within itself, a clockspring, which both strain and strife must teach us to unwind. We have seen its handiworks unravelled, like the innards of a Rolex watch, dissected on a black satin cloth in the workshop of a murdered jeweller.

'The basic unit of life is the sign, not the molecule.'
– Jesper Hoffmeyer

It is not a tangle. It is not a knot, although it might resemble a woven cable, left dishevelled, like a strand of diodes, forgotten in some bottom drawer. It is, instead, the fractal globule that unkinks itself into a wreath placed upon our tomb. We have seen it in the eddy of a whirlpool among the grottos, and we have seen it in the gyre of a whirlwind among the grasses. It is the little vortex that can torque the course of evolution for every micrococcus. It links the flinching of jellyfishes to the twinkling of dragonflies. It binds us all together via ligatures of carboxyl and amidogen. It embroiders us with error. It never regrets the wistfulness of its daydreams. It never rebukes the hellishness of its gargoyles. It is but a fuse lit long ago, its final blast delayed forever, the primacord escorting a spark through every padlock on every doorway shut against the future.

It emerges from the fluids in a bubble of montmorillonite, bursting forth, as though by fiat, to blight the entire planet. It replicates the rifling of a gun aimed at a moving target. We have seen it in the twirl of smoke from the prop wash of a biplane, tailspinning after having barrel-rolled through a dogfight. We have seen it in the contrail of a Zero, whose faithful kamikaze must loop-the-loop while he skywrites his graffiti in the clouds above his gravesite. It has printed, on the sandflat, this fragile epitaph of sigils, cursing the tsunami. It has tattooed upon itself invisible but indelible logogriphs too intricate to be utterable. It is compulsive, like a graphomaniac, unable to make his left hand stop the chalk from drawing spirals across the drywall of his cell. It is a stack of hourglasses, telling time for ballerinas who must pirouette upon their pins inside our music boxes.

It conjures forth, from nothingness, a nightingale, by reciting stray words no longer than three letters. It evokes the trilling of a songbird better than any ballad sung by choirs of sonneteers and serenaders. We have seen it in the jigsaw puzzle of a rose, whose perfect pieces lie in scattered fragments on the steps of spiral stairs. We have seen it in the ivy that, like a verdant feather boa, curls around the barberpole standing in the junkyard of our semiotic failures. It has called to mind for us a Slinky, which must somersault forever down the ascending escalator in the most sublime of all museums. It has spun the myriad raffle drums within which our lots, when chosen, summon one of us to face a sudden threat in brutal combat to the death. It is but a solenoid of copper wiring, which must embrace the iron stem of an unseen orchid, grown by electromagnets.

It is a feedback loop, feeding upon itself inside a quickening centrifuge. It is the wobble of a gyroscope, spinning inside the satellite, whose fly-by orbit slingshots a golden discus toward a distant exomoon. It burrows, like a corkscrew, through the plumes of whitewash in the wake of a torpedo. It zigzags, wayward, to our doom. It runs riot in the Von Kármán streets, where gusty winds can cause uphoisted telephone lines to whine, like sirens, in advance of a tornado. We have seen it in the twisted trusses of an extended aluminum ladder bent along its length by the ravages of a cyclone. We have seen it in the umbilicus of a waterspout, which must hula, like a stream of syrup being poured into the ocean by a storm cloud. It is but a turbofan viewed through the eyehole of a lug nut, held up, like a monocle, to the phenakistoscope of such a screw-blade.

It must build for us a giant auger that can drill a bore-hole through the azoic layer of bedrock, far below the depth of any buried fossil. It must delve through zones of Vishnu schist, far older than the ammonites now pyritized, like cogs of brass, embedded in the shale. We have seen it in the swirling flight of zebra moths succumbing to the fire, and we have seen it in the twirling plunge of sable hawks nosediving to the prey. It must plummet through a funnel, which is spinning, like a hypnodisk, at the centre of every funhouse pinwheel. It is a lathe, machining offshoots of itself, all its curlicues of shaven silver, no more than spirogyra under microscopes. It is the tusk extracted from the skull of a narwhal. It is what the fakir must evoke when he plays his ragas on a flute, bewitching a duet of vipers, curled around an ivory stick, like ribbons on a maypole.

We have seen it in the rope that hangs the felons, and we have seen it in the whip that goads the slaves. It has knit itself into a sylvan laurel, not unlike the diadem of dazzling moonlets that encompass the carousel of Saturn. It can circumnavigate a shooting star, en route to Alpha Lyræ. It can generate a gigantic field of magnetism so intense that, over time, its torsions interlace ephemeral filaments of stardust. It must crumple up the spiderweb of space-time, hauling it, like a trawl net, down into the mælstrom of a quasar. It must test itself, proving its intelligence by eternally replaying the same game of *Glasperlenspiel* upon an atomic abacus. It must calculate the odds of life delaying the doomsday of the universe. It is but a tightrope that crosses all abysses. It is but a tether that lets us undertake this spacewalk. Do not be afraid when we unbraid it.

We were never intended to be tied to whatever made us.

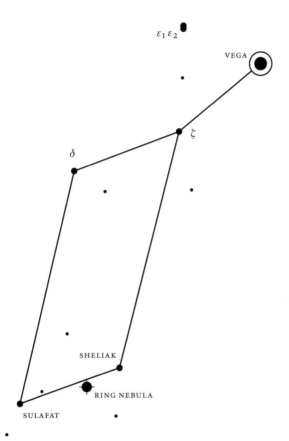

$\varepsilon_1\,\varepsilon_2$

VEGA

ζ

δ

SHELIAK

RING NEBULA

SULAFAT

THE LYRE OF ORPHEUS

VITA

EXPLICATA

 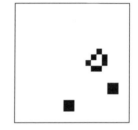

(After 57 generations.)

The Xenotext is an experiment that explores the æsthetic potential of genetics, making literal the renowned aphorism of William S. Burroughs, who claims that 'the word is now a virus.' Such an experiment strives to create a beautiful, anomalous poem, whose 'alien words' might subsist, like a harmless parasite, inside the cell of another life form. Many scientists have already encoded textual information into genetic nucleotides, thereby creating 'messages' made from DNA – messages implanted, like genes, inside cells, where such data might persist, undamaged and unaltered, through myriad cycles of mitosis, all the while saved for recovery and decoding. The study of genetics has thus granted these geneticists the power to become poets in the medium of life.

The Xenotext consists of a single sonnet (called 'Orpheus'), which, when translated into a gene and then integrated into a cell, causes the cell to 'read' this poem, interpreting it as an instruction for building a viable, benign protein – one whose sequence of amino acids encodes yet another sonnet (called 'Eurydice'). The cell becomes not only an archive for storing a poem, but also a machine for writing a poem. The gene has, to date, worked properly in *E. coli,* but the intended symbiote is *D. radiodurans* (a germ able to survive, unchanged, in even the deadliest of environments).

A poem stored in the genome of such a resilient bacterium might outlive every civilization, persisting on the planet until the very last dawn, when our star finally explodes.

Book 1 of *The Xenotext* is an 'infernal grimoire,' introducing readers to the concepts for this experiment. Book 1 is the 'orphic' volume in a diptych about both biogenesis and extinction. The book revisits the pastoral heritage of poetry, admiring the lovely idylls that rival Nature in both beauty and terror. The work offers a primer on genetics while retelling fables about the futile desire of poets to rescue love and life from the ravages of Hell. All poets pay due homage to the immortality of poetry, but few imagine that we might write poetry capable of outlasting the existence of our species, testifying to our presence on the planet long after every library has burned in the bonfires of perdition.

The Late Heavy Bombardment

Virgil welcomes everyone to the Inferno. 'The Late Heavy Bombardment' refers to the Hadean period in the history of the Earth (3.8 billion years ago), during which the world undergoes collisions from numerous meteoric impacts, all of which create a baleful, hostile environment that coincides with the genesis of all living things on the planet. The poem evokes the demonic origins of life while addressing the cosmic perils that still threaten to extinguish such a miracle.

Colony Collapse Disorder

'The Nocturne of Orpheus' is a love poem – an alexandrine sonnet in blank verse. Each line contains thirty-three letters, and together the lines form a double acrostic of the dedication; moreover, the text is a perfect anagram of the sonnet 'When I Have Fears That I May Cease to Be' by John Keats (transforming his meditation about the mortality of life into a mournful farewell by the poet before he enters Hell).

'Colony Collapse Disorder' is a 'pastoral nocturne,' which translates Book ɪᴠ of *The Georgics* by Virgil from Latin into English. Virgil addresses his patron, the Roman general Gaius Mæcenas, advising him about beekeeping, but Virgil digresses to retell two myths: first, the tale of Aristæus (who sets off to redeem his bees from Hell, expiating the murder of Eurydice through sacrifice); second, the tale of Orpheus (who sets off to rescue his love from Hell, betraying the fealty of Eurydice through misgiving). Since Virgil portrays the bees as an army, his manual almost becomes an allegory about the expiations required of a general who sends his troops to Hell but cannot bring them home. Book IV takes on special meaning for a modern reader in an era when bees are threatened with extinction. Like Aristæus, we have condemned our hives to Hell, but we remain uncertain about our ability to bring them back – and if we cannot rescue the bees, then we may be damning humanity itself to Hell, doing so when there may not be enough creatures to sacrifice so as to appease our angered deities before we all expire.

'The Xenagogue' refers to an escort who guides strangers through foreign terrain (much like Virgil, who takes a poet into Hell). Orpheus, the *xenos* (the 'foreigner'), enters the underworld, testing its hospitality, expecting the Greek edict of *xenia* (of 'offerings') to be honoured. *The Xenotext* is such an alien guest, courting the goodwill of a demonic microbe that might 'host' the poem for a future reader.

The March of the Nucleotides

'The Central Dogma' is a poetic primer, reacquainting the reader with some basic ideas in genetics. The preservation of data in DNA, via spontaneous, biochemical inscription, is still one of the most amazing wonders of the universe.

'Genetic Engineering' is a poem that revises the lyrics to a song by the same name from the album *Dazzle Ships* (1983) by the synth-group Orchestral Manoeuvres in the Dark.

'The Nucleobases' is a suite of poems derived from atomic models for the basic units of both DNA and RNA. Each text is a modular acrostic, in which the structure of a molecule defines the arrangement of a restricted vocabulary – only words of nine letters, beginning with one of the following: C (for carbon), H (for hydrogen), N (for nitrogen), or O (for oxygen). Each poem conveys a pastoral sentence about the honeybees (doing so in honour of Virgil – the first poet to have a line of poetry encoded into the genome of a flower).

'The Genetic Code' is also a poetic primer, reacquainting the reader with some basic ideas in genetics. The code is virtually universal – and surprisingly, no word used to create anything alive needs to be longer than three letters.

'The March of the Nucleotides' is a pastoral sequence of words, emulating the form of deoxyribonucleic acid (DNA). The poem enciphers the gene illustrated in 'The Enzymatic Mimeography of DNA' on page 81. A supercomputer has then converted this gene into four models for a protein.

'Death Sets a Thing Significant' is the title of a poem by Emily Dickinson, who notes that, after the death of a friend, books belonging to the departed become all the more cherished by a survivor: '[W]hen I read – I read not – / For interrupting Tears – / Obliterate the Etchings / Too Costly for Repairs.' The four 'etched' images arise from the mind of a computer, which misreads the title, interpreting it as a series of amino acids in which each letter indicates a specific molecule: D (for aspartic acid), E (for glutamic acid), A (for alanine), T (for threonine), H (for histidine), etc. The four resulting models for a protein might inspire a poet to search the canon for unplanned lipograms that, by chance, use only the twenty letters of the alphabet reserved for signifying amino acids. The poet can then convert each text into a protein to see what molecular leitmotif might result. (*The Xenotext* aspires to be among these 'little workmanships,' pondered for its meaning after the death of the author.)

'Each Orchard Claims Its Realm' responds to the first lines of poetry ever enciphered into the genome of a flower – in this case, a quote from Book II of *The Georgics* by Virgil (whose words now grow in a garden, much like one foreseen by Joe Davis, who notes that we might use actual grafts from the Tree of Knowledge to plant groves of information.)

The Virelay of the Amino Acids

'The Virelay of the Amino Acids' is a repetitious, incantatory suite of poems emulating the atomic models for each of the amino acids. The restricted vocabulary uses the constraint of an acrostic, in which each word begins with one of the following letters: c (for carbon), h (for hydrogen), n (for nitrogen), o (for oxygen), and s (for sulphur). The arrangement of words in a line corresponds to a specified structure in each molecule, and wherever this structure recurs among the molecules, so also does the line of poetry recur among the acrostics: for example, all the acids share a 'backbone' in common – a complex of amine ($-NH_2$) and carboxylic acid ($-COOH$); hence, every poem in the suite ends with the same refrain: (*no hummingbirds have / copied our opulent hymns*).

Alpha Helix

'Alpha Helix' is a delirious catalogue, listing 'manifestations' of helical imagery in the world, testifying to the ubiquity of living poetic forms by imbuing everything with the proteomic structure of life itself. The text suggests that the evolution of life may eventually play a role in the endgame of the universe, thus deciding the fate of the entire cosmos.

Acknowledgements

The Xenotext is an ongoing project that has required much perseverance – and this volume owes its completion to the devoted support of many patient friends: Nora Abrams, Andrea Andersson, Amaranth Borsuk, Bart Beaty, Braydon Beaulieu, Derek Beaulieu, Charles Bernstein, Gregory Betts, Stan Bevington, Adam Dickinson, Jeramy Dodds, Matt Donovan, Craig Dworkin, Robert Fitterman, Kenneth Goldsmith, Rohit Gupta, Bill Kennedy, Eveline Kolijn, Bronwyn Lea, Naomi Lewis, Nick Montfort, Simon Morris, Rob Oxoby, Marjorie Perloff, Sina Queyras, Hallie Siegel, Rebecca Sullivan, Nick Thurston, Tony Trehy, Kristy Trinier, Yesomi Umolu, Priscila Uppal, Peter Watts, Darren Wershler, Alana Wilcox, Patrick Wildgust, Suzanne Zelazo, *et al.*

The Xenotext also owes a debt of collegial gratitude to the following scientists who have often provided hours of professional consultation throughout the experiment: Gordon Chua, Sumukh Deshpande, Greg Fujii, Sui Huang, Stuart Kauffman, Rebecca Leong-Quong, Shawn Lewenza, Justin MacCallum, Sergei Noskov, Lisa Pimentel, Heather Rothfuss, Dennis Salahub, Raymond Turner, and Naomi Ward. (Every laboratory is an atelier for the artist of the future.)

Financial assistance for this work has been provided by the Social Science and Humanities Research Council (SSHRC), the Canada Council for the Arts, the Alberta Foundation for the Arts (AFA), the University of Calgary, and Calgary 2012. The Calgary Institute for the Humanities (CIH) has sponsored a year of my time (2013–2014) for work on this book.

Excerpts cited from 'The March of the Nucleotides' have appeared in the following venues: the magazine *Cordite Poetry Review*; the anthology *The Calgary Project: A City Map in Verse and Visual* (Frontenac House, 2014); the broadsheet *Parlour Portraits* (Book Thug, 2012); and the exhibition *Petits Genres* (Olga Korper Gallery, 2012). An excerpt from 'Colony Collapse Disorder' has appeared not only in the magazine *B After C*, but also on the website *Lemon Hound*. An excerpt from 'The Nucleobases' has appeared in the *Journal of Writing in Creative Practice*.

Each of the title pages for the chapters show two images: the leftmost one is a grid of pixels, converting the title into a QR code; the rightmost one is a grid of pixels, depicting the outcome after the QR code has been subjected to the Game of Life – an algorithm invented by the mathematician John Conway to study the behaviour of cellular automata.

Biography

Christian Bök is the author of *Eunoia* (Coach House Books, 2001), a bestselling work of experimental literature that has gone on to win the Griffin Prize for Poetic Excellence (2002). *Crystallography* (Coach House Press, 1994), his first book of poetry, has been nominated for the Gerald Lampert Memorial Award (1995). *Nature* has interviewed Bök about his work on *The Xenotext* (making him the first poet ever to appear in this famous journal of science). Bök has also exhibited artworks derived from *The Xenotext* at galleries around the world, including (among others) the Bury Art Gallery in Bury, the Olga Korper Gallery in Toronto, the Museum of Contemporary Art in Denver, the Power Plant in Toronto, and the Broad Art Museum in East Lansing. Bök teaches students of Creative Writing in the Department of English at the University of Calgary.

Typeset in Minion and Frutiger.

The Xenotext (*Book 1*) was printed at the old Coach House on bpNichol Lane in Toronto, Ontario, on Zephyr Antique Laid paper, manufactured, acid-free in Saint-Jérôme, Quebec, from second-growth forests. The text was printed with vegetable-based inks on a 1965 Heidelberg KORD offset litho press. The pages were folded on a Baumfolder, gathered by hand, bound on a Sulby Auto-Minabinda, and trimmed on a Polar single-knife cutter.

The cover depicts the poem 'Orpheus' (from *The Xenotext*), enciphered as a grey grid of cells. The image evokes the kind of results derived from the process of 'gel electrophoresis,' used to sort fragments of protein by mass.

EDITOR
Jeramy Dodds

COPYEDITOR
Alana Wilcox

DESIGNER
Christian Bök

PHOTOGRAPHER
Christian Bök

Coach House Books
80 bpNichol Lane
Toronto ON M5S 3J4
Canada

416 979 2217
800 367 6360

mail@chbooks.com
www.chbooks.com